Patrick Vennebush is a teacher's teacher as
speaker and person. His ability to weave a st
a mathematical lens is accentuated by his humor and intelligent look at life. Patrick has collected math-related jokes, one-liners, and stories and compiled them into ***Math Jokes 4 Mathy Folks,*** *and has organized them for use by educators, parents, and others who are exploring mathematics – but not taking themselves too seriously.*

His jokes and puns, although prime examples, can be humorously factored. There are the "hmm factor" thoughts that are intellectually stimulating, the "wow factor" ideas that capture real surprise, and the "clever factor" phrases that use unexpected word play.

Patrick knows that humor injects an added, spiced-up dimension to communicating mathematical ideas. He has always been generous with sharing his ideas, enthusiasm, and humor. This collection is no exception.

– L. Carey Bolster, Past President of NCSM and Recipient of the NCTM Lifetime Achievement Award, Dunedin, FL

Math Jokes 4 Mathy Folks *is a humorous compendium of mathematical jokes – some very familiar and others seemingly new. Each joke in this well-organized collection will elicit everything from a mighty groan to a knowing smile from the typical reader. It should be a valuable source of ideas for teachers to use in the first or last five minutes of a class.*

– Thomas Butts, Professor, University of Texas – Dallas

Good math jokes have two qualities. They make the audience groan, and if told right, there are sounds of crickets chirping after every punch line. The jokes in this book have both.

When teaching, humor is a great commodity. While many of us cannot think of funny math jokes, Patrick Vennebush can. In fact, he has collected them for use on your unsuspecting students. Imagine their faces when you ask them, "What do you call 3 feet of trash?" or "What did 0 say to 8?" Yes, you too can tell excellent math jokes and have your students groaning in their seats. Imagine the fun in choosing a new joke to tell every day!

– Patrick Flynn, Teacher, Olathe East High School, Olathe, KS

This book contains a lot of jokes I hadn't heard before, and I'm a very funny person. Of course, that's not necessarily a contradiction.

– Martin Funk, Teacher, New Trier High School, Winnetka, IL

Math Jokes 4 Mathy Folks is not your typical joke book. The "laugh out loud" jokes appeal to both young and old, math lovers and non-math lovers alike. This book can transform a mild-mannered mathematician into a budding comic in minutes. The cleverly written jokes incorporate mathematics concepts with humor using real-world contexts that relate to everyone. The beauty of this book is its multi-purpose appeal – the jokes in each section are arranged by level of difficulty, with jokes at the beginning of each section written to appeal to the very young, and jokes at the end of each section designed for the "after hours" crowd. Armed with this versatile text, I can imagine math teachers using jokes from this book in the classroom to help students visualize math concepts, and then using it later in the day to entertain friends, thus proving that math folks know how to have a good time, too! **Math Jokes 4 Mathy Folks** is sure to earn a treasured spot in the joke book collection of many math lovers.*

– **Latrenda Knighten,** Elementary Mathematics Coach,
East Baton Rouge Parish School District

This book is so much fun! It's a great source of jokes you can use to liven up the classroom, entertain friends, or just give yourself a laugh. It feels so good to "get" them!

– **Elizabeth Marquez,** Mathematics Assessment Specialist,
Educational Testing Service, and 1992 Recipient of the Presidential Award for Excellence in Mathematics and Science Teaching, Milltown, NJ

Math jokes come in essentially three varieties: those that make you groan, those that make you howl, and those that you don't figure out until a week after you heard them. Patrick Vennebush's comprehensive collection contains all three, of course – but it's up to you to figure out which is which!

– **Derrick Niederman**, Author of *Number Freak*
and *What the Numbers Say*, Needham, MA

Finally, a math joke book for adults! The jokes in **Math Jokes 4 Mathy Folks** *not only strike the funny bone of professional mathematicians, they also straddle that strange line between bawdy and nerdy. The witty jokes and humorous stories can be used to help students understand topics in statistics, calculus, number theory, and more. As a middle school mathematics teacher and math book collector, I find Patrick Vennebush's book refreshingly sophisticated in its silliness.*

– **Kimberly Morrow-Leong,** Math Resource Teacher,
Prince William County (VA) Schools

__Math Jokes 4 Mathy Folks__ by Patrick Vennebush is a must-read for every mathematics teacher. I'm personally delighted with this collection, because I run out of one-liners about two-thirds of the way through every semester.

I include a joke in my lesson every day, usually at the end of class. I do this to help dispel the myth of the humorless mathematician. In fact, most mathematicians have a keenly developed sense of humor. The reason is that for most mathematicians, technical words have a specific meaning that is sometimes quite different from the colloquial meaning, and often that distinction can be emphasized in a humorous way. Take the word "dawn." Two meanings come to mind, a "brainstorm" and "the start of a new day." So, it makes sense to try to get both meanings into a one-liner: "When I woke up this morning, it was dark; but then it dawned on me."

This book will make a great gift for students or a contest prize. Students might be inspired to learn more math in order to understand the more difficult jokes at the end of each section. But most everyone can appreciate the vast majority of jokes in this book.

– Harold Reiter, Professor, University of North Carolina - Charlotte

This book is an absolute gem for anyone dedicated to seeing mathematical ideas through puns, double meanings, and blatant "bad" jokes. Such perspectives help all of us see concepts and ideas in different and creative ways. You will enjoy every page!

– Jim Rubillo, Professor Emeritus,
Bucks County Community College, Newtown, PA

Let L(x) = the number of times you'll laugh, where x is the number of jokes you read from __Math Jokes 4 Mathy Folks__. Then $\lim_{x \to \infty} L(x) = \infty$.

Math jokes are a great way to reduce student anxiety about math and to help students enjoy coming to math class every day. This book has appropriate jokes for all age groups, so when I need to spice up a dull lesson and keep students' attention, I can quickly turn to an appropriate section and have plenty of jokes at the ready! Every math teacher will benefit from having this book on his or her desk.

– Jason Slowbe, Teacher, San Marcos High School, San Marcos, CA

I started reading your book to [my wife] each morning as she's primping. We were in stitches from page 1, and amazingly, I was able to follow 87% of the jokes I read (so far), although that statement may only be 59% true. All I know for sure is that there are two kinds of people in the world — and I'm one of them!

- **Joel Gray**, Digital Artist and Owner, G3D, Inc., Colorado Springs, CO

Math Jokes *is 117 pages of pure (vs. applied) fun. While I've heard a number of the jokes already, there were plenty of new ones to give me a chuckle. There are light bulb jokes, chicken jokes, one-liners, puns, and other assorted (and sordid) humor. There's something for everyone (who's a math geek or loves one).*

- **Sol Lederman**, Author, *Wild About Math*

If you appreciate clever or silly maths jokes (and who doesn't?), then this is the book for you. There are one-liners, visual gags, puns and stories, which will appeal to youngsters right up to university graduates. To start or finish a lesson, to mystify, or to amuse, this is also a perfect subject prize or competition gift.

- Australian Association of Mathematics Teachers

Math Jokes 4 Mathy Folks *is a very comprehensive collection of math-related jokes that all 'mathy' people will definitely enjoy, and math teachers could use this book to enliven their lessons. I have seen several of the jokes on the Internet, but never such a large collection.*

- **Maria Miller**, Author, *Homeschool Math Newsletter*

Teachers in particular will enjoy and want to use this book, of course, but it will provide lots of laughs for anyone else whose work involves working the numbers. It is also proof they can be very funny, too.

- **Alan Caruba**, bookviews.com

Math Jokes 4 Mathy Folks *is a treasure-trove of math-related jokes with a huge range of material, from wordplay to lightbulb jokes to visual gags to longer stories, from elementary to graduate level mathematics, from corny to subtle. Truly there's something here for everyone!*

- **Brent Yorgey**, *The Math Less Traveled*

MATH JOKES 4 MATHY FOLKS

G. Patrick
Vennebush

Robert D. Reed Publishers ▪ Bandon, OR

Robert D. Reed Publishers
P.O. Box 1992
Bandon, OR 97411
Phone: 541-347-9882; Fax: -9883
E-mail: 4bobreed@msn.com
Website: www.rdrpublishers.com

Editor: Nadine Block
Cover Designer: Cleone L. Reed
Book Designer: Debby Gwaltney

ISBN 13: 978-1-934759-48-6
ISBN 10: 1-934759-48-1

Library of Congress Control Number: 2010924043

Manufactured, Typeset, and Printed in the United States of America

Contents

Introduction

At the pub, Alex and Eli no longer talked about sampling distributions, hypothesis testing, and curve-fitting. The conversations had changed since their wives started joining them. Tonight, for instance, Kate and Susan were discussing gynecologists — a popular topic since Kate had gotten pregnant.

Eli sat quietly as the women chatted. Alex, however, longed for the days when they'd talk about Kaplan-Meier estimation and share some jokes. He couldn't take any more, so he decided to interject.

"I know a gynecologist," Alex interrupted. "She used to be a statistician. Her specialty is histerectograms."

Eli held back a smile. He wanted to laugh, but he knew that doing so would cause Kate to glare at him the way Susan was glaring at Alex. Sinking under the weight of the stare, Alex excused himself to the restroom. Eli followed.

At the urinals, Eli said to Alex, "I called our minister yesterday to let him know that Kate is having twins."

"And what'd he say?"

"He said that I should bring 'em down to the church to have them baptized. But I told him that I only wanted to baptize one of them — I'd like to keep the other as a control."

If you ask me, the jokes that Alex and Eli told are both good jokes. In fact, the one about histerectograms is a *great* joke. Of course, that's a matter of opinion. One person's knee-slapper is another person's eye-roller. The *quality* of a joke, however, may not be as important as the *appropriateness* of a joke.

All of the jokes in this book are funny. But some of them will be funnier to elementary students than to adults (those involving bananas, for instance), while others should never, under any circumstances, be told to kids under the age of 10. Some are appropriate for a high school classroom, while others should only be told at the pub. I therefore offer these jokes with a caveat, which I claim to be the "golden rule" of joke-telling: *know your audience.*

Alex would have done well to remember this rule. Math jokes about gynecologists are rarely funny to women. Eli, on the other hand, deftly handled the situation by saving his joke for the bathroom, where the line for appropriateness is slightly more difficult to discern.

To help with finding jokes appropriate for your audience, this book is divided into sections (Questions and Answers, One-Liners, etc.), and each section is arranged by "level of difficulty." In other words, the first joke within each section will be understood by the youngest math students, whereas the last joke in each section may only be intelligible to those with advanced degrees. Therefore, if you're invited to speak to your 8-year-old's class for Career Day, choose a joke from the beginning of a section; but if you're trying to break the ice at a math department cocktail party, you may be better served by telling a joke from near the end of a section.

Finally, in the interest of full disclosure, I should tell you that most jokes in this book are not original. Some of them are, like the one about Eskimos on page 19, but most have been acquired via oral tradition. I heard some from professors, and I collected others from fellow faculty members at a gifted summer camp. One was told to me by a seven-year-old cousin, and another I discovered above a urinal at a rest stop on the New Jersey Turnpike. (Seriously. I'm not making that up.)

I use these jokes often when I give a presentation to math teachers. Inevitably, one of the teachers in attendance will approach me during the first break and say, "Hey, have you heard the one about...?" Usually I haven't, and so my collection grows.

That in mind, I invite you to contribute. If you've got a math joke worth sharing, send it to **mj4mf@verizon.net**. I'll post those that I receive on my web site at **www.mathjokes4mathyfolks.com**, and if I ever feel ambitious enough to revise this book, I'll include them in the next edition.

So with that, enjoy. Brew yourself a cup of strong tea, find a comfy chair, and lose yourself in some good jokes. And may you laugh as much while reading this book as I did while putting it together.

G. P. V.
January 2010

Questions and Answers

Why is 6 afraid of 7?
Because 7 8 9.

What did 0 say to 8?
"Nice belt!"

How do you make 7 even?
Take away the *s*.

What do you get when you cross a pigeon and a zero?
A flying none.

What do you call 3 feet of trash?
A junk yard.

Which months have 28 days?
All of them.

What's the difference between a new nickel and an old dime?
Five cents.

How much do pirates pay for corn?

A buccaneer.

What insect is good with numbers?

An account-ant.

Who invented fractions?

Henry the Eighth.

What did 2 say to the other prime numbers?

"I'm glad that one of us is even-tempered!"

How many tents will a campground hold?

Ten — because ten tents make up one whole!

How much does it cost to buy sixty female pigs and forty male deer?

One hundred sows and bucks.

Why did the yardstick insist that his parents stop supporting him financially?

Because he wanted to stand on his own three feet.

How can you tell that the fractions x/c, y/c, and z/c live in a foreign country?

Because they're all over c's.

How can you tell if a mathematician is an extrovert?

When talking to you, he looks at *your* shoes.

What do you call a two-headed canary?

A binary.

A mathematician proved a famous theorem and typed up the results. But when he printed out his work, the printer failed to print the periods. How did he fix this problem?

He had his dotters do it.

Why did Mortimer refuse to drill 288 holes?

A_1: Because it was two gross.

A_2: Because it was a boring job.

What's the sine of 40?

Saying things like, "When I was your age…"

What type of lingerie does a math mermaid wear?

An algae-bra.

Why was the math book sad?

Because it had so many problems.

How can you tell that Harvard was planned by a mathematician?

The div school is next to the grad school.

What does the PhD in math with a job say to the PhD in math without a job?

"Paper or plastic?"

What did the liberal arts PhD say to the mathematics PhD?

"Would you like fries with that, sir?"

What's the difference between a math PhD and a large pizza?

A large pizza can feed a family of four.

What keeps a square in one place?

Square roots.

What is the difference between an argument and a proof?

An argument will convince a reasonable man, but a proof is needed to convince an unreasonable one.

Although horses may be able to add, why shouldn't you try to teach them analytic geometry?

You shouldn't put Descartes before the horse.

What do circles and beaches have in common?

Both have tan gents.

Why did the exterminating company hire a programmer?

They heard he was good at debugging.

Where do they put mathematicians who commit crimes?

Prism… so they can be with other convex.

What's the only cure for a bad case of right angles?

The Pythagorean serum.

How is a banquet for barbershop quartets like the series 1/2 + 1/3 + 1/4 + 1/5 + …?

Both are harmonic functions.

What did the terminating decimal say to π?

"Don't be so irrational!"

What did the statistics teacher say to console his failing student?

"Look on the bright side—you're in the top 90% of the class!"

What is the volume of a disk with radius z and height a?

pi $\cdot z \cdot z \cdot a$

What do you get if you add a rabbit, half a rabbit, one-fourth of a rabbit, one-eighth of a rabbit, and so on?

Theoretically, two rabbits – but practically, it's not possible unless you split hares.

Have you heard the latest statistics joke?

Probably…

Can an English major learn trigonometry?

Cosecant!

How do you save a drowning statistician?

Stop holding his head underwater.

What do you get if you divide the circumference of a jack-o-lantern by its diameter?

Pumpkin pi.

What do you get if you divide the circumference of the moon by its diameter?

Pi in the sky.

What do you get if you divide the circumference of an igloo by its diameter?

Eskimo pi.

How does a mathematician reprimand his children?

"If I've told you n times, I've told you $n + 1$ times…"

How are mathematicians like the Air Force?

They both use π lots (pilots).

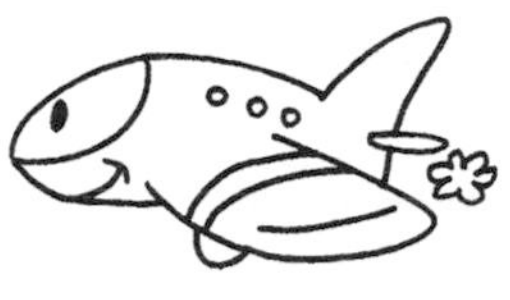

What's the difference between an economist and a confused old man with Alzheimer's?

The economist has a calculator.

What did the geometry teacher say to the algebra teacher?

"You need to stop being so symbol-minded!"

What do you call a one-sided bar with topless dancers?

A Möbius strip club.

What does a mathematician use to measure the weight of trees?

A log scale.

What do you call the matching shirt and pants of an Eskimo?

Polar coordinates.

What is non-orientable and lives in the ocean?

Möbius Dick.

Did you hear about the statistician who took a Toastmasters course?

He improved his confidence from .95 to .99.

Why don't statisticians like to model new clothes?

Lack of fit.

What does a logician pick when faced with a choice between ham salad and eternal bliss?

Ham salad — because nothing is better than eternal bliss, and ham salad is better than nothing.

Did you hear about the gynecologist who made a career change to statistics?

Her specialty was histerectograms.

What math is discussed between seabirds?

Inter-gull calculus.

What's yellow and differentiable?

A bananalytic function.

What do you call the largest accumulation point of poles?

Warsaw!

What is the value of the contour integral around Western Europe?

Zero.

Why?

Because all poles are in Eastern Europe!

Why didn't Newton discover group theory?

Because he wasn't Abel.

What did the mathematician say to his love on Valentine's Day?

A_1: My love for you is monotonically increasing over time.

A_2: When I'm away from you, my heart is like { }.

A_3: I'd like to be your derivative, so I can be tangent to your curves.

What is normed, complete, and yellow?

A Bananach space.

Why are mathematicians afraid to drive a car?

Because the width of the road is negligible compared to its length.

What is big, gray, and has integer coefficients?

An elephantine equation.

How do you make one burn?

Differentiate a log fire!

Do you know any anagrams of Banach-Tarski?

Banach-Tarski Banach-Tarski …

What is sour, yellow, and equivalent to the axiom of choice?

Zorn's lemon.

What is a mathematician's shortest joke?

Let $\varepsilon < 0$.

What's purple and commutes?

An abelian grape.

How do you insult a mathematician?

Tell him, "Your brain is smaller than any $\varepsilon > 0$!"

Why do mathematicians often confuse Christmas and Halloween?

Because Oct 31 = Dec 25.

What is green and homeomorphic to the open unit interval?

The real lime.

How is a dog like Cauchy?

It leaves a residue at every pole.

Why do truncated Maclaurin series fit the original function so well?

Because they are Taylor made.

What do you get when you cross an elephant and a banana?

$|\text{elephant}| \times |\text{banana}| \times \sin\theta$

What do you get if you cross a mosquito with a mountain climber?

You can't do it, because you can't cross a vector with a scaler (scalar).

How do you tell one bathroom full of statisticians from another?

Check the *p*-value.

Did you hear about the statistician who was thrown in jail?

He now has zero degrees of freedom.

The Light Bulb

How many mathematicians does it take to change a light bulb?

A_1: None. It's left to the reader as an exercise.

A_2: None. A mathematician can't change a light bulb, although he can easily prove that it can be changed.

A_3: One. He gives it to four programmers, thereby reducing it to a problem that was previously solved.

A_4: The answer is obvious.

A_5: Just one, once you've managed to present the problem in terms with which he is familiar.

A_6: From a previous result, it has been shown that one mathematician can change a light bulb. Additionally, if k mathematicians can change a light bulb, and if one more simply watches them do it, then $k + 1$ mathematicians will have changed the light bulb. Therefore, by induction, any number of mathematicians can change a light bulb.

How many high school math teachers does it take to change a light bulb?

Cha-a-a-a-a-a-ange?

How many math professors does it take to change a light bulb?

Just one. But he needs the help of six research students, three programmers, two post-docs, and a secretary.

How many university math lecturers does it take to change a light bulb?

Four. One to do it, and three to co-author the paper.

How many math graduate students does it take to change a light bulb?

Just one... but it takes nine years.

How many math department administrators does it take to change a light bulb?

What was wrong with the old light bulb?

How many classical geometers does it take to change a light bulb?

None. It can't be done with a straight edge and compass.

How many constructivist mathematicians does it take to change a light bulb?

None. They don't believe in infinitesimal rotations.

How many simulationists does it take to change a light bulb?

Infinity. Each one builds a fully validated model, but the light never actually goes on.

How many analysts does it take to change a light bulb?

Three. One to prove existence, one to prove uniqueness, and one to derive a non-constructive algorithm to do it.

How many number theorists does it take to change a light bulb?

No one knows the exact number, but it is believed to be an elegant prime.

How many mathematical logicians does it take to change a light bulb?

None. They can't do it, but they can prove that it can be done.

How many numerical analysts does it take to change a light bulb?

3.9967, after six iterations.

How many topologists does it take to change a light bulb?
Just one. But what will you do with the doughnut?

How many statisticians does it take to change a light bulb?
1-3, $\alpha = .05$

How many light bulbs does it take to change a light bulb?
Just one, if it knows its Gödel number.

The Chicken

Why did the chicken cross the road?

Archimedes: Give me a place to stand, and I can move the chicken across the road.

Georg Cantor: The chicken crossed the road diagonally.

Augustin Cauchy: Once the chicken crosses the road, he will integrate well with those on the other side.

Albert Einstein: Do not worry about your troubles with chickens; I assure you that mine are greater.

Paul Erdös: It was forced to do so by the chicken-hole principle.

Pierre de Fermat: I have found an admirable explanation of this, but the margin is too narrow to contain it.

Evariste Galois: He did it the night before.

Kurt Gödel: It cannot be proven that the chicken crossed the road.

David Hilbert: Because the chicken was pretty spaced out.

Isaac Newton: If chickens have been able to cross the road, it is because they have stood on the shoulders of giants.

Georg Friedrich Riemann: The answer is in Dirichlet's lectures.

Alan Turing: The chicken started to cross the road... but he may not have finished.

Xeno of Elea: He went halfway, then half again, then half yet again... but the chicken never got to the other side.

Why did the chicken cross the Möbius strip?

A_1: To get to the same side.

A_2: To get to the other… um…

Why did the chicken cross from the second to the third dimension?

The second dimension was square.

The Number

What is 666?

The number of the beast.

What is 668?

The number of the beast's next-door neighbor.

What is 666.00000?

The high-precision number of the beast. And 0.666 is the millibeast.

What is 665.95?

The retail price of the beast, and $699.25 is the price with sales tax, though you can get the beast at Wal-Mart for just $606.66.

What is 670?

The approximate number of the beast.

What is DCLXVI?

The Roman numeral of the beast.

What is 1010011010?

The binary number of the beast, and 666*i* is the imaginary number of the beast.

What is 665.999999524?

The number of the beast according to an Intel Pentium processor. The beast prepares documents in Word 6.66, his phone has area code 666, and his zip code is 00666.

Where does the beast buy gasoline?

At Phillips 666, and he loves to drive his BMW 666i along Route 666.

What will the beast do when he retires?

He'll live off the funds from his 666k retirement plan.

One-Liners

Five out of four people have trouble with fractions.

A bicycle can't stand alone because it is two-tired.

A hungry clock goes back four seconds.

A calendar's days are numbered.

Old mathematicians never die; they just lose some of their functions.

Old statisticians never die; they just undergo a transformation.

The man whose best buddy is an abacus has a friend he can count on.

A Pythagorean tree has square roots – but what kind of tree has cube roots?

Jake told Jennifer a parable about decimals – but she didn't get the point!

A recent report stated, "Given the current rate of taxation, the average American works 3½ hours every day for the government – which is 1½ more hours than the average government employee!"

Pie charts are appropriate at a pastry convention, but bar charts are never acceptable at an alcoholic support group.

She was only a mathematician's daughter, but she sure knew how to multiply.

Statistics show that those who celebrate more birthdays live longer.

A statistician can have his head in an oven and his feet in ice, and he will say that on average he feels fine.

Dear IRS: I haven't been able to sleep since I cheated on my tax return. I understated my income, so I've enclosed a check for $200. If I still can't sleep, I'll send the rest.

According to my calculations, the problem does not exist.

I started out with nothing — and I still have most of it left.

Without geometry, life is pointless.

Math problems? Call $1-800-\left[-e^{i\pi}\left(20^5\right)+\left(\sin\left(\frac{\pi}{2}\right)\div 2^{-20}\right)\right]$.

Phone operator: "I'm sorry, the number you have dialed is imaginary. Please rotate your phone 90°, and dial the number again."

The highest moments in the life of a mathematician are the first few moments after she has proven a result, but before she finds a mistake.

There are 10 kinds of people: those who understand binary, and those who don't.

There are three kinds of mathematicians: those who can count, and those who can't.

There are two types of people: those who believe that the world can be divided into two types of people, and those who don't.

When asked what term was missing from b^2 + ____ + 1, Shakespeare said, "2*b* or -2*b*... that is the question!"

Mathematics consists of 50% formulas, 50% proofs, and 50% imagination.

Life is complex: it has both real and imaginary components.

Did you hear about the constipated mathematician who worked it out with a pencil? (It was a #2 pencil.)

To a mathematician, real life is a special case.

Law of conservation of difficulties: there is no easy way to prove a deep result.

A tragedy in mathematics is a beautiful conjecture ruined by an ugly fact.

Math is like love — a simple idea, but it can get complicated.

Algebra has become so important these days that numbers will soon only have symbolic meaning.

Use variables when you don't know what you're talking about.

The statistician's wife refused to let him play with their son's toys, because she was afraid of regression.

Love is never having to say you're sorry. Statistics is never having to say you're certain.

In God we trust. All others must supply data.

A new government survey, which took 15 years to complete and cost taxpayers over $4,000,000,000, revealed that 1/2 of the people in America make up 50% of the population.

Do you know that 69.846743% of all statistics reflect an unjustified level of precision?

According to a recent government survey, 51% of people are in the majority.

According to a different survey, 33% of people say they participate in surveys.

Did you know that half of all people are below average?

More than 83% of statistics are made up on the spot.

Did you hear about the mathematician who loved his wife so much that he almost told her?

Having one wife is monogamy. Having two wives is bigotry. Having three wives is trigonometry!

Underwater ship builders are concerned with suboptimization.

The Lipton Company is big on statistics — especially *t*-tests.

There is no truth to the rumor that statisticians are mean, though they are slightly deviant, and their perspectives are skew.

Economists have forecast ten of the last six recessions.

It's impossible to have a rational conversation with a man who doesn't know his asymptote from a hole in the graph.

I heard that parallel lines actually do meet, but they are very discrete.

Some say the pope is the greatest cardinal. But others insist this cannot be so, as every pope has a successor.

Classifying mathematical problems as linear and non-linear is like classifying the Universe as bananas and non-bananas.

The most important statistic for car manufacturers is autocorrelation.

My old car is like e^x. You can derive it all you want, but you're not going to get anywhere.

You might be a mathematician if...

...you are fascinated by the equation $e^{i\pi} + 1 = 0$.

...you have memorized the first 50 digits of π.

...you know at least 10 proofs of the Pythagorean theorem.

...your telephone number is the sum of two prime numbers.

...you understand why that last one is funny.

...you believe that the World Series diverges.

...you are sure that differential equations are useful.

...you compliment your wife by mentioning that her hair is "nice and parallel."

...you say to a car dealer, "I'll take the red car or the blue one," but then you feel the need to add, "but not both."

Is it the Chinese Remainder Theorem that allows a patron to ask for a doggy bag at a Chinese restaurant?

The law of the excluded middle either rules or does not rule.

Definitions

Aftermath: the horrible headache you have after finishing an algebra test.

Arc: that which Noah built.

Bisects: how boys and girls are separated.

Catenary: a string held by a kitten at each end.

Circle: Cal, after he's been knighted.

Coincide: what you should do when it rains.

Cone: an ice-cream holder.

Corresponding Angles: angles that write letters.

Decagon: what the black jack dealer said when he lost his cards.

Defeatists: those opposed to the metric system.

Denominator: one who puts forth a candidate.

Dilemma: A lemma with two results.

Differential Operator: a distinctive person who works for the phone company.

Discrete Number: a digit with great modesty.

Divisor: what to wear to keep da sun out of your eyes.

Geometry: what an acorn says when it becomes an oak.

Golden Rule: never trust any result proven after midnight.

Inscribe: the note taker at a hotel.

Inverse: how a mathematical poet writes.

Line: the king of the jungle.

Linear Pair: a straight fruit.

Math Professor: a person who talks in other people's sleep.

Mathematician: a machine for turning coffee into theorems.

Matrices: what mathematicians sleep on.

Parabola: two neck ties worn by gauchos.

Paradox: two MD's.

Polar Bear: a rectangular bear after a coordinate transformation.

Polygon: what a pirate says when he loses his parrot.

Polynomial Ring: what a mathematician proposes with.

Prime Rib: the first derivative of a cow.

Proof: One-half percent of alcohol.

Protractor: in favor of farm machinery.

Rectangle: an angle that's been in an accident.

Secant: why the girl didn't run a 4-minute mile.

Statistician: 1) one who is good with numbers but doesn't have the personality to be an accountant; 2) a mathematician broken down by age and sex.

Tangent: a sunburned man.

Topologist: a mathematician who can't tell a coffee cup from a doughnut.

Unit: what you do with needles and yarn.

Zenophobia: the irrational fear of convergent sequences.

Slice of Pizza π

This sine has threee errors.

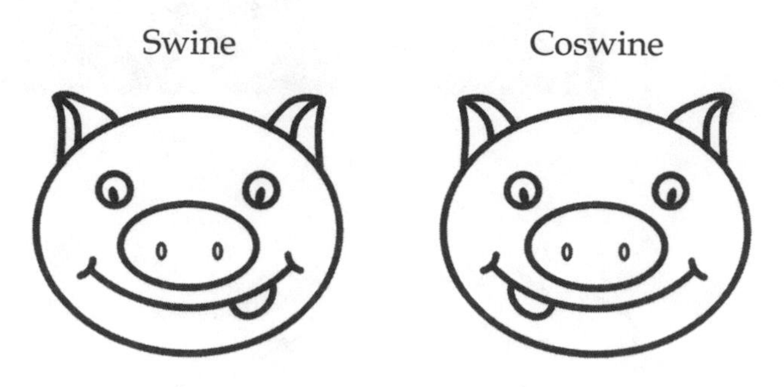

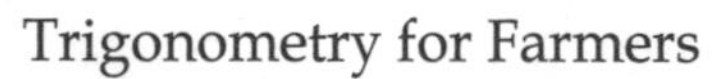
Trigonometry for Farmers

Right Wing Radical

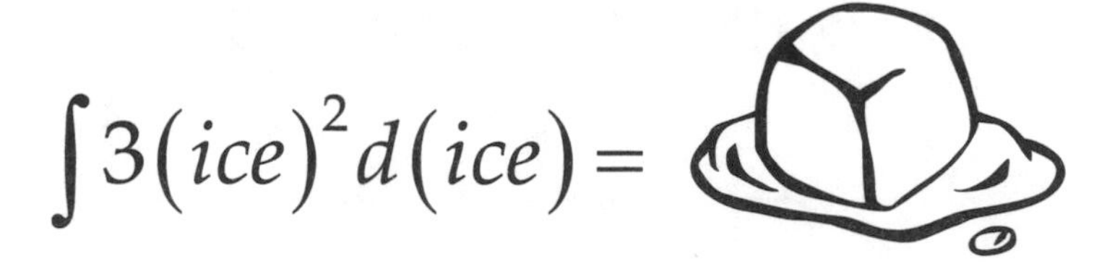

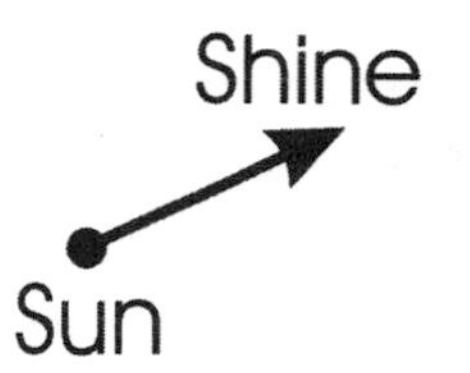

Little Ray of Sunshine

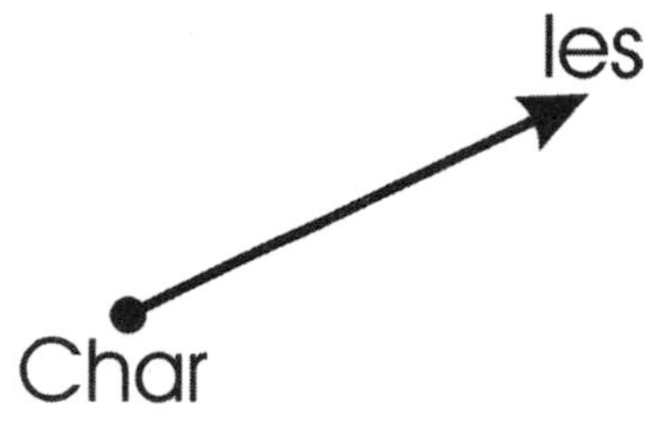

Ray Charles

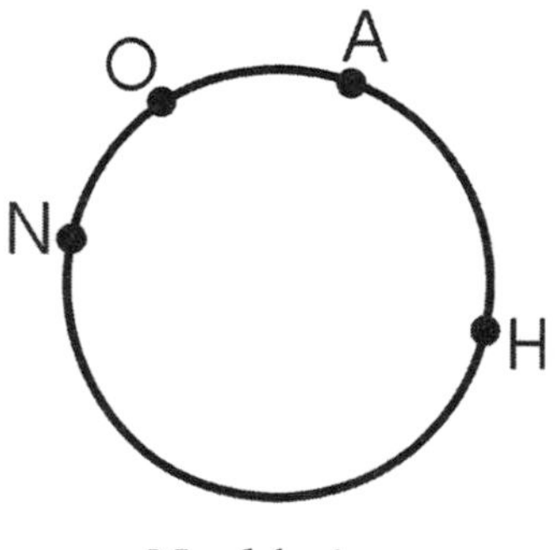

Noah's Arc

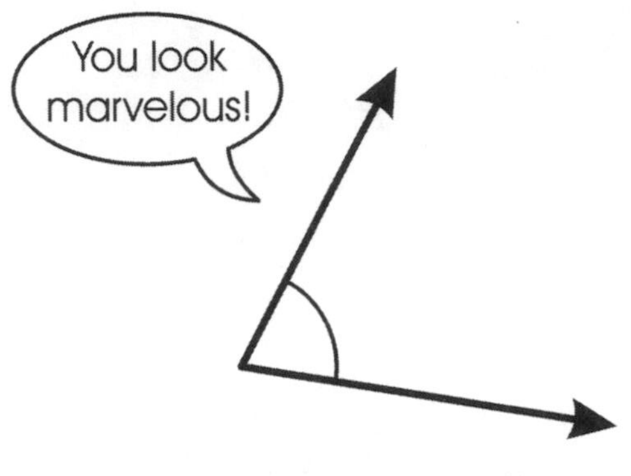

Complimentary Angle

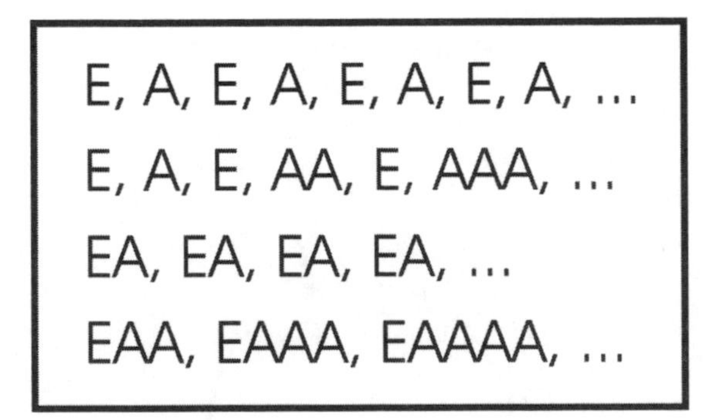

Four EA Series

Circular Reasoning

What did 0 say to 8?

Cube Roots

A pastor, a doctor, and a mathematician were stuck behind a slow foursome while playing golf. The greens keeper noticed their frustration and explained to them, "The slow group ahead of you is a bunch of blind firemen. They lost their sight saving our clubhouse from a fire last year, so we always let them play for free."

The pastor responded, "That's terrible! I'll say a prayer for them."

The doctor said, "I'll contact my ophthalmologist friends and see if there isn't something that can be done for them."

And the mathematician asked, "Why can't these guys play at night?"

A physicist, a mathematician and a computer scientist discuss whether a wife or a girlfriend is better.

The physicist claims, "A girlfriend. You still have freedom to experiment."

The mathematician replies, "A wife. You have security."

The computer scientist disagrees. "Both. When I'm not with my wife, she thinks I'm with my girlfriend. When I'm not with my girlfriend, she thinks I'm with my wife. So, I can be with my computer without anyone disturbing me!"

An engineer, a physicist, and a mathematician are subjects in a psychology experiment. After not eating for 24 hours, each of them is then locked in a room with a can of food, but without an opener, and each of them has some paper and pencil.

At the end of the experiment, the psychologists enter the engineer's room. The pencil and paper are unused, but there are dents all over the walls. The engineer is eating from the open can – he threw it against the walls until it split open.

They enter the physicist's room next. The paper is covered with formulas, there is one dent in the wall, and the physicist is eating, too – he calculated how to throw the can against the wall, and he got it open on the first try.

In the mathematician's room, the paper is also full of formulas, the can is still closed, and the mathematician has disappeared. But there are strange noises coming from inside the can. They use an opener to open the can, and the mathematician jumps out. "Damn!" he says. "I got a sign wrong…"

A mechanical engineer, an electrical engineer, and a computer engineer are riding in a car when it breaks down. They pull off to the side of the road.

"Pop the hood," says the mechanical engineer. "I'm sure it's a problem with the gears, so I'll go take a look."

"No," says the electrical engineer. "I'm sure it's a problem with the wires. Pop the hood, and I'll go take a look."

"I've got a better idea," says the computer engineer. "Why don't we just get out, get back in, and try to start it again?"

A mathematician, a physicist, and an engineer are given identical rubber balls and told to find the volume.

The mathematician pulls out a measuring tape and records the circumference. He then divides by 2π to find the radius, cubes that, multiplies by π again, and finally multiplies by 4/3, thereby calculating the volume.

The physicist gets a bucket of water, places 1.00000 gallons of water in the bucket, drops in the ball, and measures the displacement to six significant figures.

The engineer writes down the serial number of the ball and looks it up.

An investment company is hiring mathematicians. During their interviews, three recent graduates – a pure mathematician, an applied mathematician, and a graduate in mathematical finance – are asked what starting salary they expect.

Timidly, the pure mathematician asks, "Would $30,000 be too much?"

The applied mathematician says confidently, "I think $60,000 would be fair."

And the math finance graduate asks flippantly, "How about $300,000?"

Astounded, the personnel officer says, "Are you kidding me? We have a graduate in pure mathematics willing to do the same work for a tenth of what you are demanding!"

"I accounted for that," says the math finance grad. "I allocated $135,000 for me, $135,000 for you, and $30,000 to pay the pure mathematician to do the work."

A biologist, a physicist and a mathematician were sitting at a street cafe. Across the street, a man and a woman entered a building. A few minutes later, the same man and woman exited the building with another person.

"Ah," says the biologist. "They have multiplied!"

"No," sighs the physicist. "It is an error in measurement."

The mathematician says, "If exactly one person enters now, the building will be empty."

An engineer, a chemist and a mathematician are staying in three adjoining cabins at an old motel. First the engineer's coffee maker catches fire. He smells the smoke, wakes up, unplugs the coffee maker, throws it out the window, and goes back to sleep.

Later that night, the chemist smells smoke, too. He wakes up and sees that a cigarette butt has set the trash can on fire. He says to himself, "Hmm. How does one put out a fire? One can reduce the temperature of the fuel below the flash point, isolate the burning material from oxygen, or both. This could be accomplished by applying water." So he picks up the trash can, puts it in the shower stall, turns on the water, and, when the fire is out, goes back to sleep.

The mathematician, of course, has been watching all this out the window. So later, when he finds that his pipe ashes have set the bed sheet on fire, he is not in the least taken aback. He says, "Aha! A solution exists!" and goes back to sleep.

A mathematician, scientist, and engineer are each asked, "Suppose we define a horse's tail to be a leg. How many legs does a horse have?"

The mathematician answers, "5."

The scientist answers, "1."

And the engineer objects, "But you can't do that!"

A farmer enlisted the help of an engineer, a physicist, and a mathematician to help him fence off the largest possible area with a given amount of fence.

The engineer made a circle and proclaimed that he had the most efficient design.

The physicist made a long, straight line and proclaimed, "We can assume the length is infinite," and pointed out that fencing off half of the Earth was certainly a more efficient solution.

The mathematician laughed. "No, no, no," he said, and he built a very tiny fence around himself. "I declare myself to be on the outside."

A chemist, physicist, and mathematician are stranded on a desert island when a can of food washes ashore. The chemist and physicist devise many ingenious ways to open the can. Then suddenly the mathematician gets a bright idea. "Let's assume we have a can opener…"

A physicist, a statistician, and a mathematician place a bet on a horse race.

The physicist's horse comes in last. "I don't understand," he says. "I determined each horse's strength through a series of careful measurements."

The statistician's horse comes in next to last. "How is this possible?" he asks. "I evaluated the results of all races for the past month."

They both look at the mathematician, whose horse came in first. "How did you do it?"

"Well," he explains. "First, I assumed that all horses were identical and spherical..."

A biologist, a physicist, a mathematician and a computer scientist were on vacation in Scotland. Glancing from a train window, they saw a black sheep in the middle of a field.

"How interesting," observed the biologist. "The sheep in Scotland are black!"

"No, no!" the physicist corrected. "Some Scottish sheep are black!"

The mathematician elaborated further. "In Scotland, there exists at least one field, containing at least one sheep, at least one side of which is black."

And the computer scientist said, "Oh, no! A special case!"

Same Question, Different Answers

Three recent graduates – a pure mathematician, an applied mathematician, and a statistician – are being interviewed by the Department of Defense. All three are asked the same question, "What is one-third plus two-thirds?"

The pure mathematician says firmly, "One."

The applied mathematician takes out his pocket calculator, punches in the numbers, and replies, "0.99999999."

The statistician asks, "That depends. What would you like it to be?"

The scientist asks, "Why does it work?"

The engineer asks, "How does it work?"

The accountant asks, "How much will it cost?"

The philosopher asks, "Do you want fries with that?"

Several people are asked, "What is π?"

The engineer says, "Pi is approximately 22/7."

The physicist says, "Pi is 3.14159."

The mathematician says, "Pi is the ratio of circumference to diameter."

And the nutritionist says, "Pie should be eaten in moderation!"

Several people were asked, "What is 2 × 2?"

An engineer whips out his graphing calculator, presses a bunch of buttons, and finally announces, "3.99, plus or minus 0.01."

The physicist consults his technical references, sets up the problem on his computer, and announces, "Somewhere between 3.98 and 4.02."

The mathematician thinks for a moment then declares, "I don't know the answer — but I have a proof that an answer exists!"

The philosopher smiles and asks, "What do you mean by 2 × 2?"

The logician replies, "You'll need to define 2 × 2 more precisely."

The sociologist says, "I don't know, but it sure was nice talking about it."

A behavioral ecologist concludes, "A polygamous mating system."

Finally, a medical student says plainly, "2 × 2 is 4."

The others look at him, astonished. When asked how he knew that, he replies, "I memorized it."

A mathematician was put in a room with a table and three metal spheres. He was told to do whatever he wants with the balls and the table. After an hour, the balls are arranged in a triangle at the center of the table.

The same test is given to a physicist. After an hour, the balls are stacked one on top of the other in the center of the table.

Finally, an engineer is given the test. After an hour, one of the balls is broken, one is missing, and he's carrying the third out in his lunchbox.

An engineer, a physicist and a mathematician were asked to hammer a nail into a wall.

The engineer built a Universal Automatic Nailer, a device able to hammer every possible nail into every possible wall.

The physicist conducted several experiments, and she developed a new technology of hammering nails at super-low temperatures.

The mathematician generalized the problem to an n-dimensional problem of penetration of a knotted one-dimensional nail into an $(n - 1)$-dimensional hyper-wall. Several fundamental theorems are proved. Of course, the problem is too rich to suggest a simple solution, and even the existence of a solution is far from obvious.

Several people were asked to test the following hypothesis: *All odd numbers greater than 1 are prime.*

A mathematician reasons, "3 is prime, 5 is prime, 7 is prime, but 9 is not prime. Therefore, the hypothesis is false."

A math grad student says, "3 is prime, 5 is prime, 7 is prime, and – by induction – every odd integer greater than 1 is prime."

A physicist says, "3 is prime, 5 is prime, 7 is prime, 9 is not prime, 11 is prime, and 13 is prime. Hence, five out of six experiments support the hypothesis. It must be true."

Another physicist argues, "3 is prime, 5 is prime, 7 is prime, 9 is an experimental error, 11 is prime. Just to be sure, try several randomly chosen numbers: 17 is prime, 23 is prime, ..."

An engineer says, "3 is prime, 5 is prime, 7 is prime, 9 is approximately prime, 11 is prime, ..."

A computer science student, reading the output from the screen, says, "3 is prime, 3 is prime, 3 is prime, 3 is prime, ..."

A biologist says, "3 is prime, 5 is prime, 7 is prime, 9 – more research is needed."

A psychologist claims, "3 is prime, 5 is prime, 7 is prime, 9 is prime but tries to suppress it, ..."

A chemist asks, "What's a prime?"

A politician states, "Some numbers are prime, but the goal is to create a kinder, gentler society where all numbers are prime."

A programmer says, "Wait a minute, I think I have an algorithm for finding prime numbers... give me just a minute, I think I've found the last bug... no, that's not it... ya know, I think there may be a compiler issue here – oh, wait, did you want IEEE-998.0334 rounding or not? – was that in the specs? – hold on, I've almost got it – I was up all night working on this program, ya know... if management would just get me that new workstation that I've been asking for, I'd be done by now..."

Puns

Mathematical puns are the first sine of dementia.

Salesman: Lady, this vacuum cleaner will cut your work in half.

Woman: Great. I'll take two of them.

A lazy dog is a slow pup.

A slope up is an inclined plane.

An ink-lined plane is a sheet of writing paper.

Therefore, a lazy dog is a sheet of writing paper.

For a good prime, call 555-7523.

My wife shops at the deci-mall. When she parks, she puts money in the deci-meter.

Graphing rational functions is a pain in the asymptote.

Lumberjacks make good musicians because of their natural logarithms.

The flood is over and the ark has landed. Noah lets all the animals out and says, "Go forth and multiply."

A few months later, Noah decides to see how the animals are doing. Everywhere he looks, he finds baby animals. Everyone is doing fine, except for one pair of snakes. "What's the problem?" asks Noah.

"Cut down some trees, and let us live there," say the snakes.

Though confused by the request, Noah does as they wish. Several weeks pass. Noah checks on the snakes again. There are lots of little snakes. "This is wonderful," Noah says, "but how did the trees help?"

"We're adders," say the snakes, "and we need logs to multiply."

An abacus is one tool you can always count on.

During a transatlantic flight, both the pilot and co-pilot became ill. Very worried, the flight attendants asked if there was anyone aboard who could fly the plane. An elderly Polish man raised his hand. "I used to fly planes in WWII," he said. "I'll give it a shot."

He was escorted to the cockpit, and he sat in the pilot's seat. But after several minutes, he still was looking over the equipment, and one of the flight attendants finally hollered at him, "C'mon! Can you fly this plane or not?"

The old man calmly replied, "Please, have patience. I'm just a simple Pole in a complex plane."

After having dinner together, one mathematician asked the other, "How was your meal?"

"Great, but $\frac{\sqrt{-1}}{8}$."

"Yeah," said the other, "i over 8, too."

As easy as 3.141592653...

Möbius strippers only show you their back side.

Russell to Whitehead: "My Gödel is killing me!"

There were three medieval kingdoms on the shores of a lake. For years, the kingdoms had been fighting over an island in the middle of the lake. Finally, the three kings decided that they would send their knights out to do battle, and the winner would take the island.

The night before the battle, the knights and their squires pitched camp and readied themselves. The first kingdom had twelve knights, and each knight had five squires, all of whom were busily polishing armor, brushing horses, and cooking food. The second kingdom had seventeen knights, and each knight had eight squires. Everyone at that camp was also busy preparing for battle. At the camp of the third kingdom, however, there was only one knight with just one squire. This squire took a large pot and hung it from a looped rope in a tall tree. He busied himself preparing the meal, while the knight polished his own armor.

When the hour of battle came, the three kingdoms sent their squires out to fight.

The battle raged, and when the dust had cleared, the only person left was the lone squire from the third kingdom, having defeated the squires from the other two kingdoms, thus proving that the squire of the high pot and noose is equal to the sum of the squires of the other two sides.

One day, Jesus said to his disciples: "The Kingdom of Heaven is like $3x^2 + 8x - 9$."

A man who had just joined the disciples looked very confused and asked Peter, "What does he mean by that?"

Peter replied, "Don't worry – it's just another one of his parabolas."

Conversion Chart

10^{21} piccolos = 10^{9} los = 1 gigolo

10^{18} acts = 1 exact

10^{15} shops = 1 petashop

10^{12} pins = 1 terrapin

10^{12} bulls = 1 terabull

10^{12} Italian firms = 1 terra firma

10^{12} microphones = 1 megaphone

10^{9} antics = 1 gigantic

10^{6} bicycles = 2 megacycles

1 trillion pins = 1 terrapin

1,000,000 aches = 1 megahurtz

2000 pounds of Chinese soup = Won ton

2000 mockingbirds = 2 kilomockingbirds

1000 manjaros = 1 kilomanjaro

1000 milliliters of wet socks = 1 liter hosen

500 millinaries = 1 seminary

454 graham crackers = 1 pound cake

Time it takes to sail 220 yards at 1 nautical mile per hour
= 1 Knot-furlong

365.25 days of drinking low-calorie beer = 1 lite year

100 rations = 1 C-ration

16.5 feet in the Twilight Zone = 1 Rod Serling

16 ounces Alpo = 1 dog pound

10 cards = 1 decacards

10 rations = 1 decoration

10 monologues = 5 dialogues

10 millipedes = 1 centipede

8 nickels = 2 paradigms

8 catfish = 1 octopus

2 monograms = 1 diagram

2 snake eyes = 1 paradise

2 wharves = 1 paradox

2 physicians = 1 paradox

4 chutes = 2 parachutes

007 secret agents = 1 mole

25 cagey bees = 1 mole

3 1/3 tridents = 10 dents = 1 decadent

2.4 statute miles of medical tubing at Yale = 1 I.V. League

1 kilogram of falling figs = 1 fig newton

1/2 large intestine = 1 semicolon

1/2 lavatory = 1 demijohn

10^{-1} bell = 1 decibel

10^{-1} mate = 1 decimate

10^{-2} mental journeys = 1 centimental journey

10^{-3} tarries = 1 military

10^{-6} fish = 1 microfiche

10^{-6} mouthwash = 1 microscope

10^{-6} phones = 1 microphone

10^{-9} Nanettes = 1 nanoNanette

10^{-12} boos = 1 picoboo

10^{-12} de gallo = 1 pico de gallo

10^{-15} bismol = 1 femtobismol

10^{-18} boys = 1 attoboy

Weight of a televangelist = 1 billigram

Speed of a tortoise breaking the sound barrier = 1 Mach turtle

Basic unit of laryngitis = 1 hoarsepower

Shortest distance between two jokes = A straight line

Time between slipping on a peel and hitting the pavement
= 1 bananosecond

Professions

"My life is all basic arithmetic," a young businesswoman explains. "I try to add to my income, subtract from my weight, divide my time, and avoid multiplying."

The chef instructs his apprentice, "Take two-thirds water, one-third cream, one-third broth, ..."

The apprentice interrupts, "But that's already four-thirds!"

"So, get a larger pot!"

Engineers think that equations approximate the real world.

Scientists think that the real world approximates equations.

Mathematicians are unable to make the connection.

A statistician and an economist are drowning. You can only save one of them. What do you do – go for a jog, or grab a sandwich?

A cannibal is buying brains for dinner. In the butcher's shop, he is told that mathematician brain costs $1 a pound, engineer brain costs $2 a pound, and politician brain costs $4 a pound.

The cannibal is bewildered. He asks the butcher, "Why does politician brain cost four times as much as mathematician brain? Is the quality that much better?"

The butcher replies, "No, but it takes a lot more politicians to make a pound."

During his daily briefing, the secretary of defense informed the president, "Yesterday, three Brazilian soldiers were killed."

"Oh, no!" the president exclaimed. "That's terrible!" He dropped his head into his hands, and he sat silent for a while. Finally, he looked up and asked, "How many is a brazillion, anyway?"

A woman goes to the doctor's office, and the doctor tells her she has only six weeks left to live.

"Doctor, that's terrible! What should I do?" she asks.

"Are you married?" he asks.

"No."

"Then I suggest you find yourself an actuary and marry him," the doctor advises.

"Will that help me live longer?" she asks.

"Well, no... but it'll *feel* longer."

Three statisticians are duck hunting. A duck passes overhead. The first statistician shoots and misses six inches too high. The second shoots and misses six inches too low. And the third shouts, "We got it! We got it!"

A mathematician and an engineer are on a deserted island. They find two palm trees with a coconut in each one. The engineer climbs up one tree, gets the coconut, and eats it. The mathematician climbs up the other tree, gets the coconut, then climbs the first tree, and places the coconut in it.

"What'd you do that for?" asks the engineer.

The mathematician replies, "I've now reduced it to a previously solved problem."

A consulting statistician and his client sat down together for the first time.

The client explains, "I desperately need your help interpreting the significant three-way interaction in this factorial ANOVA. What are your fees?"

The statistician says, "One-hundred dollars for three questions."

The client asks, "Isn't that a little steep?"

The statistician replies, "Not at all! Now, what's your third question?"

Some engineers are trying to measure the height of a flagpole. They only have a measuring tape, and they have not been able to slide the tape up the pole. A mathematician asks what they are doing, and they explain.

"The solution is easy," he says. He pulls the pole out of the ground, lays it down, and measures it.

After he leaves, one of the engineers says, "That is so typical! We tell a mathematician that we need to know the height — and he gives us the length!"

A quiet little man was brought before a judge. The judge looked down at the man and then at the charges and then down at the little man again. "Can you tell me what happened?" he asked the man.

"I'm a mathematical logician, dealing in the nature of proof."

"Yes, go on," said the judge.

"Well, I was at the library, and I found the books I wanted and went to take them out. The librarian told me I had to fill out a form to get a library card, so I filled out the forms and got back in line."

"And?" said the judge.

"And the librarian asked, 'Can you prove you're from New York City?' So I stabbed her."

As a statistician passes the security checkpoint, they discover a bomb in his carry-on bag. He is detained for questioning.

"Sir, I must admit that I don't understand!" says the security officer. "Why would a man of your status try to bring a bomb on a plane?"

"Well," says the statistician, "data shows that the probability of a bomb being on an airplane is 1 in 1000. That's too great a risk for me to feel comfortable. But the probability of there being two bombs on a plane is 1 in 1,000,000. Therefore, if I bring a bomb on the plane, the chance of there being another bomb on the plane is highly unlikely."

A patient asks his surgeon what the odds are of him surviving an impending operation. The doctor replies that the odds are 50-50. "But there's no need to worry," the doctor explains. "The first fifty have already died!"

A statistician's wife has twins. He was delighted, and he called to tell his minister the good news.

"Excellent!" said the minister. "Bring them to church on Sunday, and we'll baptize them."

"No," replied the statistician. "Let's just baptize one. We'll keep the other as a control."

A mathematician decides that he is sick of math, and he tells the fire chief he wants to be a fireman.

The fire chief says to him, "You seem like a good guy, and I'd be glad to hire you, but first I have to give you a little test."

The chief takes the mathematician to the alley behind the fire department which contains a dumpster, a spigot, and a hose. The chief says, "Okay, you walk into the alley, and the dumpster is on fire. What do you do?"

The mathematician replies, "Well, I connect the hose to the spigot, turn the water on, and put out the fire."

The chief says, "That's great, perfect. Now, what would you do if you're walking down the alley, and you see that the dumpster is not on fire?"

The mathematician thinks for a while. Finally, he says, "I light the dumpster on fire!"

The chief yells, "What? That's horrible! Why would you light the dumpster on fire?"

The mathematician replies, "Well, that way I reduce the problem to one I've already solved."

The statistician was asked by his friend why he always used the urinal on the far end.

He replied, "Because there is only half the probability of being sprayed by someone else."

A mathematical biologist is hiking, and he encounters a shepherd with a large herd of sheep.

"How much for one of your sheep?" he asks the shepherd.

"Sorry, they aren't for sale," the shepherd replies.

The math biologist thinks for a moment and then says, "I will tell you the precise number of sheep in your herd without counting. If I'm right, don't you think that I deserve one of them as a reward?" The shepherd shrugs, but agrees. The math biologist says, "387."

The shepherd is amazed. "You're right. I hate to lose one of my sheep, but I promised. Take one."

The math biologist grabs one of the animals and puts it on his shoulders. He is about to walk away when the shepherd says, "Wait! If I can tell you what your profession is, don't you think I deserve to get my animal back as a reward?"

"That's fair enough."

"You must be a mathematical biologist."

The man is stunned. "You're right. But how did you know?"

"It was easy. You gave me the precise number of sheep without counting — but then you picked up my dog."

A logician on his way to his favorite fishing hole saw a sign that read, "All the worms you want for $1.00." He stopped and ordered $2.00 worth.

Two statisticians were flying on a plane. An hour into the flight, the pilot announces that they have lost an engine, but they shouldn't worry because there are still three left. However, instead of 3 hours, the trip will now take 4 hours.

A little later, the pilot announces that a second engine has failed, but they still have two left, only now it will take 6 hours to get to their destination.

Later still, the pilot announces that a third engine has gone out. Never fear, he announces, because the plane can still fly safely with just one engine. However, the trip will now take 12 hours.

At this point, one statistician turns to the other and says, "I sure hope we don't lose that last engine, or we'll be up here forever!"

A math professor and his wife pack all their belongings into cardboard boxes and have them shipped off to a new home. The professor heads to the new house a few days before his wife, and he calls her when he arrives.

"All 39 boxes have arrived," the professor tells his wife.

"But there should be 40 boxes," says the wife.

The professor counts again, but he once again gets 39.

Furious, the wife calls the moving company and complains, but they tell her that 40 boxes were delivered.

She calls her husband back. "I don't understand," she says. "When you count, you get 39, but they counted 40. That's very odd."

"Well," the professor says, "Stay on the line and count with me. Zero, one, two, three, ..."

A math professor and a minister reach the Pearly Gates, and St. Peter starts them on a tour.

They walk for a little while and stop in front of a palatial mansion. There are fountains and ponds in the front yard, huge columns on the front porch. It's absolutely gorgeous. Both the minister and professor are speechless. St. Peter turns to the math professor and says, "This is your house, where you'll be spending all of eternity."

St. Peter tells the minister to follow him, and beyond the mansion, they pass a mid-sized home, a smaller suburban house, and a mobile home. Finally, they reach a tree house that appears to have been built from leftover scrap wood. St. Peter points at the tree house and says to the minister, "This is the house in which you'll spend eternity."

The minister is flabbergasted. "The math professor gets a mansion," he says incredulously, "and all I get is a tree house?"

"We base the size of your home on how many people you got to pray during your years on Earth. And a whole lot more people prayed in his math class than ever prayed in your church!"

When the logician's son refused to eat his vegetables, the father threatened him, "If you don't eat your veggies, you won't get any ice cream!" The son, frightened at the prospect of not having his favorite dessert, quickly finished his vegetables. After dinner, impressed that his son had eaten all his vegetables, the father sent his son to bed without any ice cream.

Two men have been drifting in a hot air balloon for hours, hopelessly lost. They have been floating through a cloudbank. Suddenly, the clouds part, and they notice a man standing on a nearby mountain.

"Hello, down there! Can you tell us where we are?"

The man on the mountain thinks for a moment, then just as the balloon is about to re-enter the clouds, he shouts, "You're in a balloon!"

"That must have been a mathematician," says one of the men in the balloon.

"Why?"

"Because he thought long and hard about what to say, and his response was absolutely correct, but it was completely useless."

A mathematician and an engineer attend a lecture by a physicist. The topic concerns processes that occur in spaces with dimensions of 9, 12, and higher. The mathematician is enjoying the lecture immensely, but the engineer is terribly confused. By the end, the engineer has a headache.

After the lecture, the engineer asks the mathematician, "How do you understand all this stuff?"

The mathematician responds, "I just visualize the process."

"How can you possibly visualize something that occurs in 9-dimensional space?"

"Oh, that's easy," says the mathematician. "First, I visualize it in n-dimensional space, then I let $n = 9$."

A group of mathematicians and a group of engineers are traveling by train to attend a conference. Each engineer has a ticket, but only one of the mathematicians has a ticket. Of course, the engineers are ridiculing the mathematicians, when suddenly one of the mathematicians yells, "Conductor coming!"

All of the mathematicians disappear into one restroom.

The conductor checks the ticket of each engineer and then knocks at the restroom door. "Ticket, please."

The mathematicians slide their one ticket under the door. The conductor checks it and leaves. A few minutes later, when the coast is clear, the mathematicians come out of the restroom. The engineers are very impressed.

When the conference ends, the engineers decide that they are as smart as the mathematicians, so they buy just one ticket. But this time, the mathematicians get on board with no ticket. As the engineers start to question them, one of the mathematicians shouts, "Conductor coming!"

The engineers rush to the restroom. A few seconds later, one of the mathematicians knocks at the restroom door and says, "Ticket, please."

A logician at the grocery store is asked, "Paper or plastic?"

He responds, "Not 'not paper and not plastic'!"

For a psychology experiment, a hungry mathematician and a hungry physicist are seated in chairs, and a hot meal is placed on a table on the opposite side of the room. The psychologist explains, "You must stay in your chairs. Each minute, I will move your chair half the distance toward the table."

The mathematician snarls, "That's ridiculous! You know I'll never reach the food!" And he storms out of the room.

The physicist, however, starts salivating. The psychologist is a little confused.

"Don't you realize that you'll never reach the food?" he asks.

The physicist replies, "Of course — but I'll get close enough for all practical purposes!"

A physics professor has been conducting experiments and has worked out a set of equations that seem to explain his data. Nevertheless, he is unsure if his equations are really correct, and he asks a colleague from the math department to check them.

A week later, the math professor calls him. "I'm sorry, but your equations are complete nonsense."

The physics professor is disappointed, of course. Strangely, however, his incorrect equations turn out to be surprisingly accurate in predicting the results of further experiments. So, he asks the mathematician if he was sure about the equations being completely wrong.

"Actually," the mathematician replies, "they are not *complete* nonsense. But the only case in which they are true is the trivial one where the field is Archimedean."

"Wasn't yesterday your first anniversary? What's it like being married to a mathematician for a whole year?"

"She just filed for divorce."

"I don't believe it! Did you forget about your anniversary?"

"No. Actually, on my way home last night, I stopped at a flower shop and bought a bouquet of red roses for her. When I came home, I gave her the roses and said, 'I love you.'"

"So, what happened?"

"She slapped my face, kicked me in the groin, and threw me out of the apartment."

"That's terrible!"

"No, no, it's all my fault, actually… I should have said, 'I love you and *only* you.'"

A newlywed husband is discouraged by his wife's obsession with mathematics. Afraid of being second fiddle to her profession, he finally confronts her. "Do you love math more than you love me?"

"Of course not, dear — I love you much more!"

Happy, although skeptical, he challenges her. "Then prove it!"

She thinks for a bit, and then says, "Okay. Let epsilon be greater than zero…"

"That math professor's marriage is falling apart!"

"No wonder! He's into scientific computing – and she's incalculable!"

A mathematician asked a fortune teller, "Tell me, are the proofs to unsolved theorems found in heaven?"

"I have good news, and I have bad news," she replied.

"What's the good news?"

"Not only are all of the proofs revealed in heaven, but you will be shown the most elegant proofs possible!"

"That's awesome! What's the bad news?"

"By this time tomorrow, you'll have an elegant proof of the Riemann hypothesis."

Pure Math

Seventeen asked 6 and 28, "Don't you two ever do anything wrong?"

"Nope," they said. "We're perfect!"

"Statistics shows that most people are abnormal!"

"How's that?"

"According to statistics, the average person has one breast and one testicle..."

A mathematician stumbled through the door at 3am. His wife was furious and yelled, "You're late! You said you'd be home by 11:45!"

The mathematician replied, "No, honey, I said I'd be home by a quarter of twelve."

A devout Christian asks a mathematician, "Do you believe in one God?"

"Yes," replies the mathematician. "Up to isomorphism!"

A mathematician at the history museum asked a tour guide, "How old is your skeleton of the T. Rex?"

"It is 60 million, 4 years old."

"How do you know with such precision?"

"Well, one of the scientists told me that the skeleton was 60 million years old when I started working here – and that was four years ago."

A mathematician organizes a raffle in which the prize is an infinite amount of money, paid over an infinite amount of time. Of course, with the promise of such a prize, his tickets sell like hot cakes.

When the winning ticket is drawn, the jubilant winner comes to claim the prize. The mathematician explains the method of payment: "1 dollar now, 1/2 dollar next week, 1/3 dollar the week after that, ..."

Aleph-null bottles of beer on the wall,

Aleph-null bottles of beer,

You take one down, pass it around,

Aleph-null bottles of beer on the wall...

The devil approaches a mathematician who has spent most of his life trying to prove the Riemann hypothesis. "In exchange for your soul," the devil tells him, "I'll give you a proof of the hypothesis." The mathematician agrees, and the devil promises to return with a proof in two weeks.

Ten days later, the mathematician can hardly control his excitement. He tells all his colleagues that he is close to a proof. He brags to his students. He distributes press releases. But at the end of two weeks, the devil has not returned. In fact, six months pass before the mathematician sees the devil again.

"Where have you been?" the mathematician asks when the devil finally shows up.

"I'm sorry, I was not able to prove the theorem," says the devil. "But I think I found some really interesting lemmas!"

Two mathematicians are studying a convergent series.

The first one says, "Do you realize that the series converges even when all the terms are made positive?"

"Are you sure?"

"Absolutely!"

A high school math teacher was arrested trying to board a flight while in possession of a compass, a protractor, and a graphing calculator. He was charged with carrying weapons of math instruction. According to law enforcement officials, he is believed to have ties to the Al-Gebra network, and sources close to the investigation say they saw him commiserating with radicals.

A mathematician gives a talk intended for a general audience. The talk is announced in the local newspaper, but he expects few people to show up because most people will not be able to make sense of the title, *Convex sets and inequalities*.

To his surprise, the auditorium is packed when the talk begins. When he finishes, someone in the audience raises his hand.

"But you said nothing about the actual topic of your talk!"

"What do you mean?"

"In the paper, this talk was listed as *Convicts, Sex, and Inequality*."

A mathematician and a stockbroker go to the racetrack. The broker is ready to bet, but the mathematician is not yet comfortable. Before he does, he wants to understand the rules and have a look at the horses.

"Don't worry," the broker says. "I know an empirical algorithm that allows me to find the number of the winning horse with absolute certainty." The mathematician is unconvinced.

"You're being too theoretical!" the broker exclaims, and he places a $10,000 bet on a horse.

When the broker's horse comes in first, the mathematician is dumbfounded. "What is your algorithm?" he wants to know.

"It's rather easy. I have two children, three and five years old. I add up their ages, and I bet on that number."

"But three plus five is eight – and you bet on horse number nine!"

"I told you, you're too theoretical! Didn't I just experimentally prove that my calculation is correct?"

$1 \approx 2$, for sufficiently large values of 1.

The numbers 6 and 28 are walking down the street when 6 starts having a conversation, seemingly with himself. "Who are you talking to?" asks 28.

"I'm talking to $36 + 4i$," replies 6.

"To whom?" asks 28. "There's no one there!"

"Oh, sorry," says 6. "You can't see him. He's my imaginary friend."

Axiom 1: Knowledge = Power.

Axiom 2: Time = Money.

Work = Power × Time

By substitution, Work = Knowledge × Money. Solving for Money then gives:

Money = Work ÷ Knowledge

Consequently, as Knowledge decreases, Money increases, regardless of how much Work is done. Thus, we can conclude: The less you know, the more you make.

Theorem. Every positive integer is interesting.

Proof. Assume that there is an uninteresting positive integer. Then there must be a smallest uninteresting positive integer. But being the smallest uninteresting positive integer is interesting by itself. Contradiction!

Theorem. A cat has nine tails.

Proof. No cat has eight tails. Since one cat has one more tail than no cat, it must have nine tails.

Theorem. A dollar doesn't go as far as it used to.

Proof. Note that 1 dollar = 100 cents. Then,

$$\begin{aligned} 1 \text{ dollar} &= 100 \text{ cents} \\ &= (10 \text{ cents})^2 \\ &= (0.1 \text{ dollars})^2 \\ &= 0.01 \text{ dollars} \\ &= 1 \text{ cent} \end{aligned}$$

Theorem. 4 = 3.

Proof. Suppose $a + b = c$. Then,

$$\begin{aligned} 4a - 3a + 4b - 3b &= 4c - 3c \\ 4a + 4b - 4c &= 3a + 3b - 3c \\ 4(a + b - c) &= 3(a + b - c) \\ 4 &= 3 \end{aligned}$$

Theorem. 1 = ½.

Proof. Note that

$$\begin{aligned}\frac{1}{1\cdot 3}+\frac{1}{3\cdot 5}+\frac{1}{5\cdot 7}+\cdots &= \frac{1}{2}\left(\left(\frac{1}{1}-\frac{1}{3}\right)+\left(\frac{1}{3}-\frac{1}{5}\right)+\left(\frac{1}{5}-\frac{1}{7}\right)+\cdots\right)\\ &= \frac{1}{2}\left(\frac{1}{1}+\left(-\frac{1}{3}+\frac{1}{3}\right)+\left(-\frac{1}{5}+\frac{1}{5}\right)+\left(-\frac{1}{7}+\frac{1}{7}\right)+\cdots\right)\\ &= \frac{1}{2}(1+0+0+0+\cdots)\\ &= \frac{1}{2}\end{aligned}$$

However, the same infinite series can also be rewritten as follows:

$$\begin{aligned}\frac{1}{1\cdot 3}+\frac{1}{3\cdot 5}+\frac{1}{5\cdot 7}+\cdots &= \left(\frac{1}{1}-\frac{2}{3}\right)+\left(\frac{2}{3}-\frac{3}{5}\right)+\left(\frac{3}{5}-\frac{4}{7}\right)+\cdots\\ &= \frac{1}{1}+\left(-\frac{2}{3}+\frac{2}{3}\right)+\left(-\frac{3}{5}+\frac{3}{5}\right)+\left(-\frac{4}{7}+\frac{4}{7}\right)+\cdots\\ &= 1+0+0+0+\cdots\\ &= 1\end{aligned}$$

Thus, ½ = 1.

A student walks into the math department holding a shiny, new trophy. He explains, "I won it in a math contest. They asked what 9 + 8 is. I said, '15' – and I got 3rd place!"

A college dean says to the chair of the physics department, "Why do I always have to give you guys so much money for labs and expensive equipment and stuff? Why can't you be more like the math department? All they need is money for pencils, paper, and wastebaskets. Or even better, be like the philosophy department – all they need are pencils and paper."

Professor: This term, I will offer topology.

Student: No problem. Topology accepted.

A math graduate student is riding a bike across campus when a friend stops him.

"Hey, is that a new bike?" the friend asks.

"Sure is," he says.

"Where'd you get it?"

"You know that freshman girl I've been tutoring? Well, she found out yesterday that she passed calculus. She was so excited, she biked immediately to my place, took off all her clothes, jumped on my bed, and said, 'I am so grateful. You can have anything you want!'"

"Good call on the bike," said the friend. "I don't think her clothes would have fit.

A mathematician has been invited to speak at a conference. His talk is titled, "Proof of the Riemann Hypothesis."

When the conference actually takes place, he speaks about something completely different.

After his talk, a colleague asks him, "Did you find an error in your proof?"

He replies, "Nope. I never had one."

"So why did you make this announcement?"

"That's my standard precaution – in case I die on my way to the conference."

"Students nowadays are so clueless," a math professor complains to his colleague. "Yesterday, a student came to my office and asked if General Calculus was a Roman war hero."

A medical company figured out how to package basic knowledge in pill form. A student goes to the pharmacy and asks what kinds of pills are available. The pharmacist says, "Here's a pill for English literature." The student takes the pill, swallows it, and has new knowledge about English literature.

"What else do you have?" asks the student.

"I also have pills for art history, biology, and world history," replies the pharmacist.

The student takes all three of them, and he has new knowledge about those subjects, too.

Then the student asks, "Do you have a pill for math?"

The pharmacist says, "Wait just a moment." He goes into the store room and brings back a huge pill and heaves it onto the counter.

"That's the pill for math?" the student inquires. "It must weigh five pounds!"

The pharmacist replied, "Well, you always knew that math was a little hard to swallow."

At the math department mixer, an overconfident sophomore boy approached a popular senior girl. "$y = 2x + 3$," he said.

"Oh, please," she said. "I've heard that line before."

A little later, a junior approached.

"You have the most beautiful blue $\sqrt{-1}$'s," he said.

"I bet you say that to all the girls," she replied.

A senior then asked her, "Are you a differentiable function?"

Skeptical, she asked, "Why?"

"Because I'd like to be tangent to your curves!"

A professor's enthusiasm for teaching pre-calculus varies inversely with the likelihood of his having to do it.

"Isn't statistics wonderful?"

"How so?"

"Well, according to statistics, there are 42 million alligator eggs laid every year. Of those, only about half get hatched. Of those that hatch, three-fourths are eaten by predators in the first 36 days. And of the rest, only 5 percent make it to one year of age. Isn't statistics wonderful?"

"What's so wonderful about that?"

"If it weren't for statistics, we'd be up to our asses in alligators!"

The professional quality of a mathematician is inversely proportional to the importance he attaches to space and equipment.

The relationship between pure and applied mathematicians is based on trust and understanding. In particular, pure mathematicians do not trust applied mathematicians, and applied mathematicians do not understand pure mathematicians.

Four friends who have been doing really well in calculus decide to take a weekend road trip instead of studying for the final. Not surprisingly, they drink too much, oversleep on the morning of the final, and get back to campus after the exam is over. One of them concocts an explanation and asks for the professor's mercy. "We went to my parent's cabin for the weekend so we could have a quiet place to study. Driving back on Monday morning, we got a flat tire. We didn't have a spare, and it took hours for help to arrive."

Sympathetically, the professor says, "I understand. Come back tomorrow morning, and I will allow you to take a make-up exam."

On Tuesday morning, the professor seats the students in the four corners of a large lecture hall. The tests – which are just one page – are already on the desks, and the professor tells the students to begin. On the front side of the paper is just one question involving a definite integral, and all four students answer it easily.

On the back side of the paper is an even simpler question: "Which tire?"

At a conference, a mathematician proves a theorem.

Someone in the audience interrupts him. "But, sir, that proof must be wrong. I have found a counterexample."

The speaker replies, "I don't care — I have another proof for it."

A statistics professor was completing what he thought was a very inspiring lecture on the importance of significance testing in today's world. A young nursing student in the front row sheepishly raised her hand and asked, "But, sir, why do nurses have to take statistics?"

The professor thought for a few seconds and replied, "Young lady, statistics saves lives!"

The nursing student was utterly surprised and after a short pause retorted, "But, sir, please tell us how statistics saves lives!"

"Well," the professor said angrily, "Statistics keeps idiots out of the nursing profession!"

A math student and a computer science student are jogging together in a park when they hear a voice say, "Please, help me!"

They stop to look. The voice belongs to a frog sitting in the grass.

"Please, help me!" the frog repeats. "I'm not really a frog. I'm an enchanted, beautiful princess. Kiss me, and the spell will be broken."

The computer science student picks up the frog and examines it carefully from all sides – but he doesn't kiss it.

"You won't have to marry me," the frog continues. "I'll do whatever you wish if you'll just kiss me."

The computer science student puts the frog into one of his pockets.

"But why don't you kiss her?" the math student asks.

"You know," the computer science student replies, "I just don't have time for a girlfriend right now – but a talking frog is a really cool pet!"

At the end of his course on optimization, a professor sternly looked at his students and said, "There is one final piece of advice I'm going to give you. No matter what you have learned in this course, never apply it to your personal lives!"

"Why?" the students asked.

"Well, some years ago, I observed my wife preparing breakfast, and I noticed that she wasted a lot of time walking back and forth in the kitchen. So, I attempted to optimize the procedure, and I told my wife about the results."

"And what happened?"

"Before I applied my expert knowledge, my wife needed nearly an hour to prepare breakfast for the two of us. Now, it takes me less than fifteen minutes to make my own."

These days, some mathematicians are so tense that they are no longer able to sleep during seminars.

A mother of three is pregnant with her fourth child.

One evening, the eldest daughter says to her dad, "Do you know what I learned today?"

"No."

"Our new baby will be Chinese!"

"What?"

"That's right!" said the daughter. "I read in the newspaper that every fourth child born nowadays is Chinese."

"What happened to your girlfriend, that really cute math student?"

"She's not my girlfriend any more. She was cheating on me. A couple of nights ago, I called her on the phone, and she told me that she was in bed wrestling with three unknowns."

"Why did you break up with your boyfriend, the math major?" a mother asked her daughter.

"He was terrible – restless during the days, and he couldn't sleep at night. He was always trying to solve his math problems. When he finally solved one, he wasn't happy then, either. He'd call himself a complete idiot for not solving the problem sooner, and he'd throw all his notes into the garbage. One day, I couldn't take it anymore, and I told him to drop math. And do you know what he told me?"

"What?"

"He said he enjoyed it!"

A statistics student accelerated before crossing every intersection. His passenger finally asked, "Why do you go so fast through intersections?"

The student replied, "Statistically speaking, you're far more likely to have an accident at an intersection, so I try to spend less time there."

A math major once faced the following question on an elementary physics exam: "You are given an accurate barometer. How would you use it to determine the height of a skyscraper?"

The student answered, "Go to the top floor, tie a long piece of string to the barometer, lower the barometer until it touches the ground, and then measure the length of the string."

The professor, unsatisfied with the answer, interviewed the student. "Can you give me another method, one that demonstrates your knowledge of physics?"

"Sure. Go to the top floor, drop the barometer, and measure how long it takes to hit the ground."

"That's not quite what I had in mind. Would you like to try again?"

"Okay," said the student. "Make a pendulum of the barometer, measure its period at the bottom, then measure its period at the top."

"Please try again," the professor demanded.

"Measure the length of the barometer, mount it vertically on the ground on a sunny day, and measure its shadow. Then, measure the shadow of the skyscraper."

"This is ludicrous," announced the professor. "Again."

"Walk up the stairs, and use the barometer as a ruler to measure the height of the walls in the stairwells."

"Last chance, young man."

"Find where the janitor lives, knock on his door, and say 'Please, sir — if I give you this barometer, will you tell me the height of the skyscraper?'"

A math professor is talking to her little brother who just started his first year of graduate school in mathematics.

"What's your favorite thing about mathematics?" the brother wants to know.

"Knot theory," she says.

"Yeah, me neither."

A math professor is lecturing to his class. He formulates a theorem, writes it on the chalkboard, and says, "Of course, this result is immediately obvious."

Seeing blank stares from his students, he turns toward the board to think about what he has written. He thinks for a moment, strokes his beard, scratches his head, and finally leaves the room.

He is gone for several minutes. Just when the students think he may not return, he runs back into the room and declares, "Yes, yes… indeed, it *is* obvious!"

Professor: Parallel lines meet at infinity.

Student: Infinity must be a noisy place with all those lines crashing together!

Professor: Suppose the number of sheep is x.

Student: Yes, ma'am – but what if the number of sheep is not x?

A math professor claims to his students that he can prove everything under the assumption that 1 + 1 = 1.

A student challenges him. "Then prove that you're the pope!"

He thinks for a minute and then replies, "I am one, and the pope is one. Therefore, the pope and I are one."

A mathematician decides that she wants to learn more about practical problems. She sees a seminar with an interesting title, The Theory of Gears. So she decides to attend. The speaker stands up and begins, "The theory of gears with a real number of teeth is well known…"

A statistics professor distributes a 50-question true-false test. During the exam, the professor notices that one student, sitting in the very back of the room, is flipping a coin and writing down answers. This goes on for two full hours, and after all other students have left, the student is still flipping coins and writing answers.

The professor approaches the student. "Listen," she says, "I know you weren't prepared for the test, and I know that you've been flipping a coin to determine the answers. But what I don't understand is, what could possibly be taking you so long?"

"Shhh," says the student. "I'm checking my answers!"

An absent-minded math professor was stopped by one of his graduate students on the campus quad. They chatted for a few minutes, and at the end of the conversation, the professor asks the student, "Which way was I going when you stopped me?"

"You were going that way, sir," the student says as he points.

"Excellent!" shouts the professor. "Then I've already had my lunch!"

Teacher: Who can tell me what 9×6 is?

Student: 54!

Teacher: Excellent! And who can tell me what 6×9 is?

Student: 45!

Teacher: How many seconds are in a year?

Student: Twelve. January 2nd, February 2nd, …

Son: My math teacher is crazy.

Mother: Why do you say that?

Son: Yesterday, she told us that $3 + 2$ is 5. Today, she told us that $4 + 1$ is 5!

Teacher: What is $11q - q$?

Student: $10q$.

Teacher: You're welcome.

Teacher: What is $2k + k$?

Student: 3000!

The math professor's six-year-old son knocks at the door of his father's study. The son says, "I need help with a math problem I couldn't do at school."

"Sure," the father says. "Tell me what the problem is."

"Well, it's a really hard one. There are four ducks swimming in a pond, and two more ducks join them. How many ducks are now swimming in the pond?"

The professor stares at his son with disbelief. "You couldn't do that? All you need to know is that $4 + 2 = 6$!"

"I'm not stupid!" the son says. "Of course, I know that $4 + 2 = 6$. But what does that have to do with ducks?"

A student told his math teacher, "Ms. Smith, I agreed to solve equations. I agreed to prove theorems. I even agreed to perform chi-square tests. But graphing is where I draw the line!"

The math teacher asks her students, "How can you divide fourteen sugar cubes into three cups of coffee so that each cup has an odd number of sugar cubes?"

"That's easy," says Johnny. "One, one, and twelve."

"But twelve isn't odd," replies the teacher.

"Oh, yes, it is," Johnny replies. "Twelve is a very odd number of sugar cubes to put in a cup of coffee!"

Son: Dad, I'm going to a party tonight. Can you do my math homework for me?

Dad: I'm sorry, son, it just wouldn't be right.

Son: That's okay. Can you give it a try anyway?

Young girl: Can you help me find the lowest common denominator?

Grandmother: Haven't they found that yet? They were looking for that when I was in school!

Student: The computer ate my data. I think it's trying to get me in trouble.

Teacher: Don't anthropomorphize computers – they hate that!

The instructor in a college algebra class was explaining that terms cannot be subtracted from one another unless they are like terms. "For example," she said, "you can't take five apples from six oranges."

"But," asks a student, "can't you take five apples from three trees?"

A teacher was explaining to her geometry class that it was physically impossible to trisect an angle. A young woman raised her hand and protested, "That's not true. I've seen it done."

The teacher replied, "You must be mistaken. It is physically impossible."

"Well," the girl said, "when I get to heaven, I'll just ask Euclid."

Annoyed, the teacher asked, "And what if Euclid went to hell?"

The girl replied, "Then you can ask him."

A father is concerned about his son's bad grades in math, so he decides to enroll his son in a Catholic school. After the first marking period, the son brings home his report card, and he's getting an A in math.

The father is pleased, but he asks his son, "Why are your math grades suddenly so good?"

The son explains, "I knew they meant business when I saw that guy on the wall nailed to a plus sign!"

Mother: Why does the tablecloth you just put on the table have the word 'truth' written on it?

Daughter: Because my teacher said I would need a truth table to complete my homework!

Teacher: Why isn't your homework done?

Student_1: I could have sworn I put the homework inside a Klein bottle, but this morning I couldn't find it.

Student_2: I locked the paper in my trunk, but a four-dimensional dog got in and ate it.

Student_3: I walked halfway to my textbook, then half again, then half again. But I was never able to reach it.

Student_4: I watched the World Series, then spent the rest of the night trying to prove that it converged.

Student_5: I couldn't figure out if I am the square root of -1 or if *i* is the square root of -1.

Student_6: Yesterday was overcast, and I have a solar-powered calculator.

Student_7: I did my homework, but the margin was too small to contain it.

Student_8: I divided by zero, and my paper burst into flames.

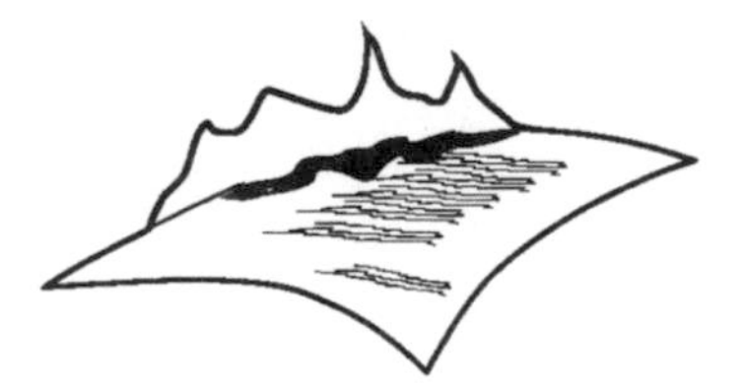

Don't Drink and Derive

The number 15 walks into a bar and orders a beer.

"I'm sorry," said the bartender. "I can't serve you."

"Why not?"

"Because you're under 21."

An old barfly tries to pick up a mathematician.

"How old do you think I am?" she asks coyly.

"Well, by the fire in your eyes, I'd say you're 18. By the glow in your cheeks, I'd say you're 19. And by the radiance in your face, I'd say you're 20. Adding that up is something you can probably do for yourself..."

Rene Descartes is in a bar. The bartender asks, "Would you like another?" Descartes replies, "I think not," and he disappears.

Two men are having a good time in a bar. Outside, there's a terrible thunderstorm. Finally, one of the men thinks that it's time to leave. Since he has drunk a lot, he decides to walk home.

"But aren't you afraid of being struck by lightning?" his friend asks.

"Not at all. Statistics shows that one person per year gets struck by lightning in this part of the country – and that person was struck by lightning three weeks ago."

Alcohol and calculus don't mix. Please don't drink and derive.

Some statisticians don't drink because they are *t*-test totalers. Others drink a lot, as evidenced by the proliferation of box-and-whiskey plots.

In topologic hell, beer is packed in Klein bottles.

A chemist, an engineer, and a mathematician stumbled across a pile of beer cans. Unfortunately, they're the old-fashioned cans that do not have the tab at the top. One of them proposed that they split up and find can openers.

The chemist went to his lab and concocted a magical chemical that dissolves the top of the can in an instant but then immediately evaporates so that the beer is not affected.

The engineer went to his workshop and created a new Hyper-Opener that can open 25 cans per second.

They went back to the pile with their inventions and found the mathematician finishing the last can of beer. "How did you do that?" they asked in astonishment.

The mathematician answered, "Oh, well, I just assumed they were open and started drinking."

The police department was told to crack down on vagrancy, so it was easy when a drunk staggered toward a cop and asked, "'Scuse me, offisher, what time ish zit?"

The cop replied, "It's one o'clock," and gave him a bonk on the head with his baton.

"My goodness," said the drunk. "I'm glad it's not midnight!"

The optimist thinks the glass is half full. The pessimist thinks the glass is half empty. And the engineer thinks the glass is twice as big as it needs to be.

An attractive female accountant was having a drink when the man next to her asked for her phone number.

She paused for a moment, and then replied, "I'm sorry, I've seen so many figures today. I just can't remember my exact telephone number – but I can probably estimate it to within 10 percent."

Two math professors are sitting in a pub.

The first one complains, "Isn't it disgusting how little the general public knows about mathematics?"

"Well," his colleague replies, "perhaps you're just too pessimistic."

"I don't think so," the first one replies. A few minutes later, the first professor excuses himself to the bathroom. When he leaves, the other professor signals to the waitress.

"When my friend comes back, I'll wave you over to our table. I'll then ask you a question, and I'd like you to answer, 'x to the third over three.' Can you do that?"

The girl giggles and repeats several times, "x to the third over three, x to the third over three, x to the third over three, ..."

When the first professor returns, his colleague says, "I was thinking about what you said. I bet the waitress knows a lot more about mathematics than you imagine."

He calls her over and asks, "Miss, can you tell us what the integral of x-squared is?"

She replies, "x to the third over three."

The first professor's mouth drops wide open, and his colleague grins smugly as the waitress adds, "...plus a constant."

A group of mathematicians had been drinking at the same pub for years. During that time, they had heard the same jokes so often that they assigned numbers to them. To save time, instead of telling an entire joke, they would just shout out its number.

"48," yelled one of them. The others all laughed heartily at this old classic.

"184," shouted a female mathematician, which elicited several guffaws.

"354," barked another, which brought only mild laughter from his friends. But one young mathematician began rolling on the floor, holding his stomach and laughing hysterically.

The teller of the last joke approached the young mathematician and asked, "What about my joke did you find so funny?"

"I hadn't heard that one before."

The same group of mathematicians was again at the pub when a graduate student joined them.

After a few jokes had been told, the grad student decided to take part. He shouted, "555!" No one laughed, but a few of the others scowled at him.

He leaned over to one of the older mathematicians and asked what was wrong.

"Young man, there are ladies present, and we don't tell dirty jokes in mixed company."

Once again, the same group of mathematicians was sitting around and telling jokes.

After a few jokes had been told, a young mathematician shouted, "86!" No one laughed. Then some of the other mathematicians shouted numbers, and people laughed again.

Later, the young mathematician asked one of the others why no one had laughed at his joke.

"It's all in how you tell it, friend."

Two homeless guys were sitting at a bar, lamenting their lot. "I'm so poor," one of them said.

Just then, Bill Gates walked into the bar.

"Cheer up," said his friend. "On average, everyone in this bar just became a billionaire."

A mathematician walks into a bar and orders 1/2 of a beer. Another mathematician walks in and orders 1/4 of a beer. Another walks in and orders 1/8 of a beer. Then 1/16, 1/32, 1/64, 1/128, and so on. The bartender looks at them, says, "You're all idiots," and proceeds to pour them one beer.

Graduate students at the local university had just designed the Robot Bartender 3000. The RB3K can make any drink as well as tailor the conversation to your intellect. It was placed in a local tavern for testing.

A woman walked into the bar and the RB3K asked her, "What's your IQ?"

"150," she said. So the robot proceeded to talk to her about dynamical systems, mathematical physics, and operations research.

Another woman walked into the bar. "What's your IQ?" asked the robot.

"130." So RB3K struck up a conversation about differential calculus, discrete mathematics, and topology.

A third woman then walked in. "What's your IQ?"

"80," she said. "Why?"

"Oh, no reason," said RB3K. "So, how are things in the English department?"